MW00451748

SCIENCE NOTEBOOK

This notebook belongs to:

____

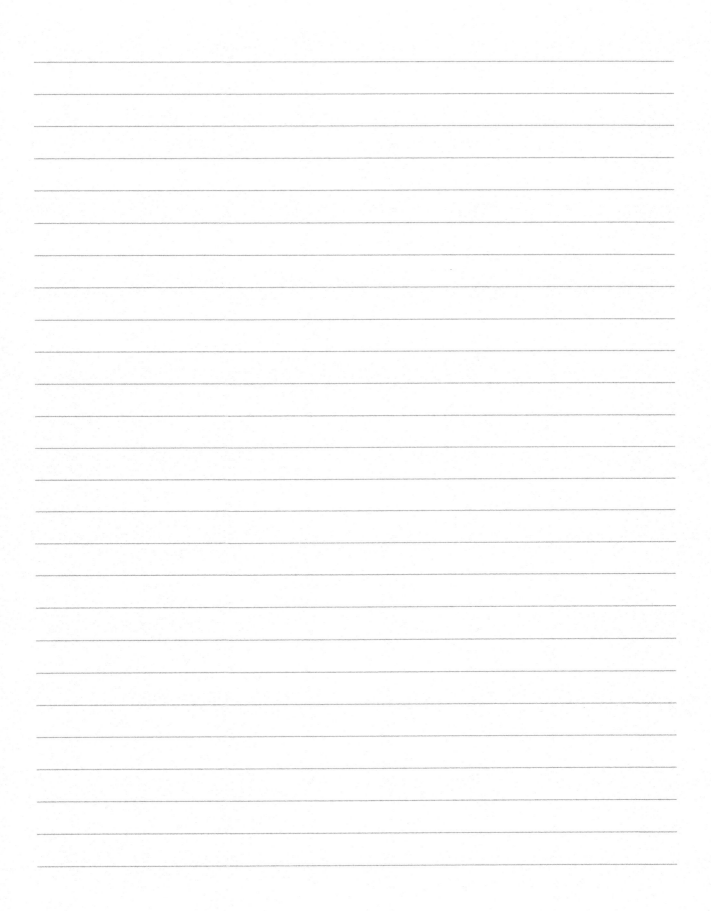

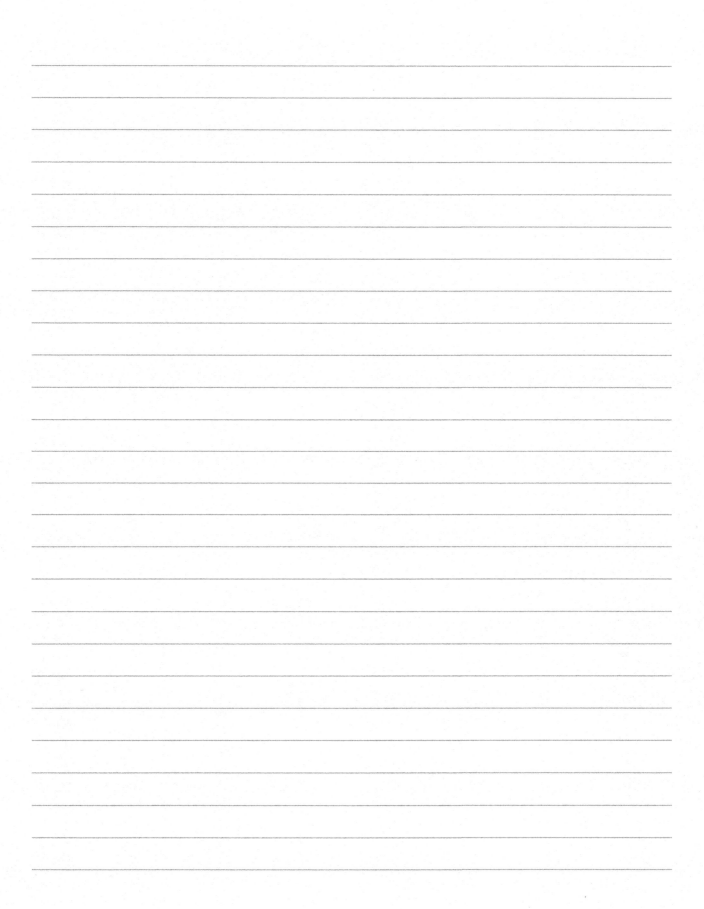

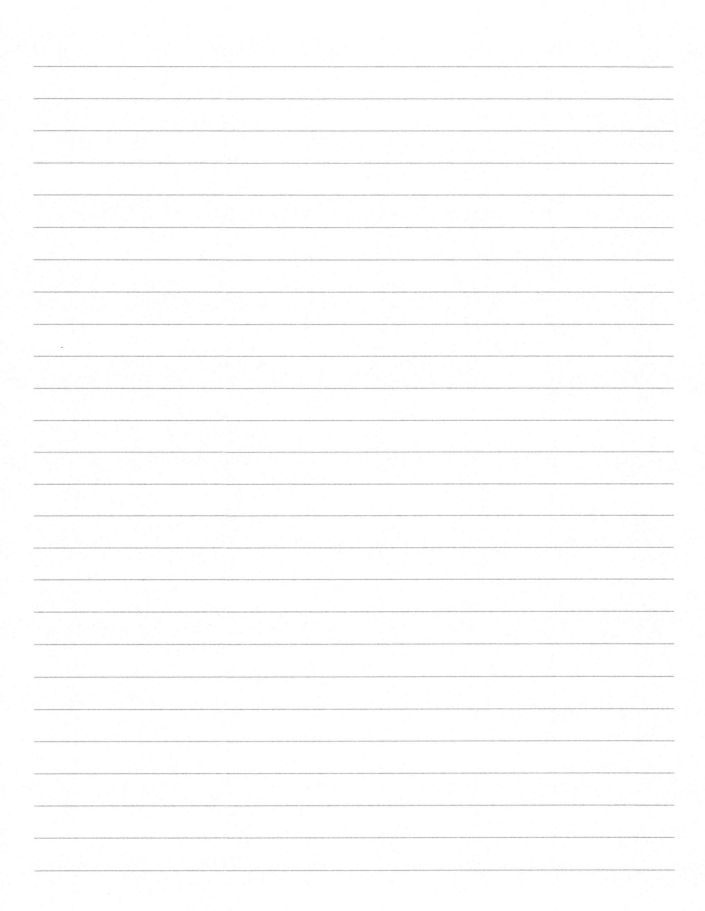

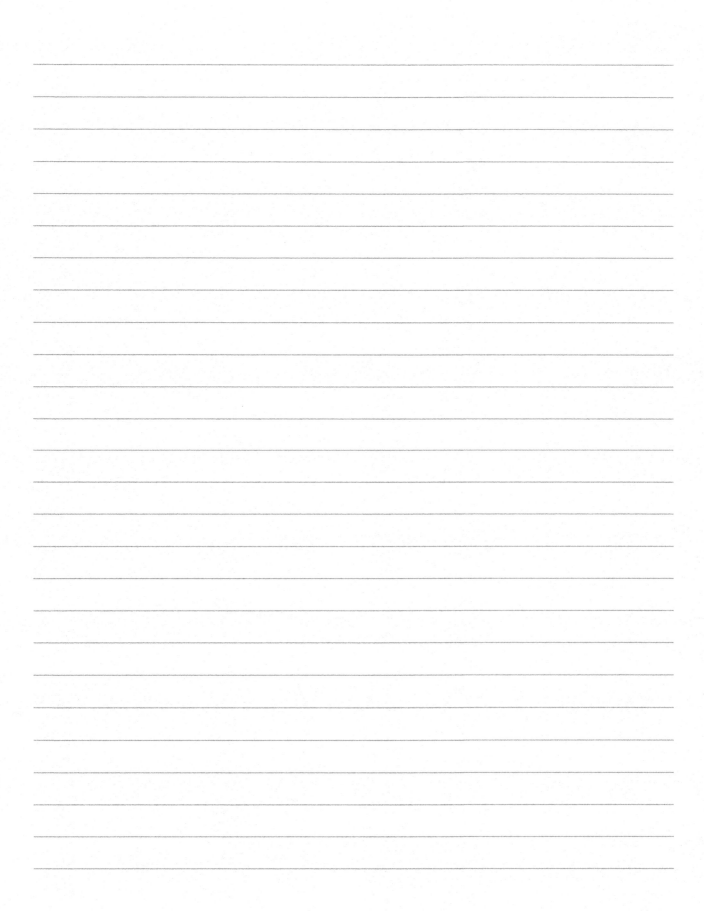

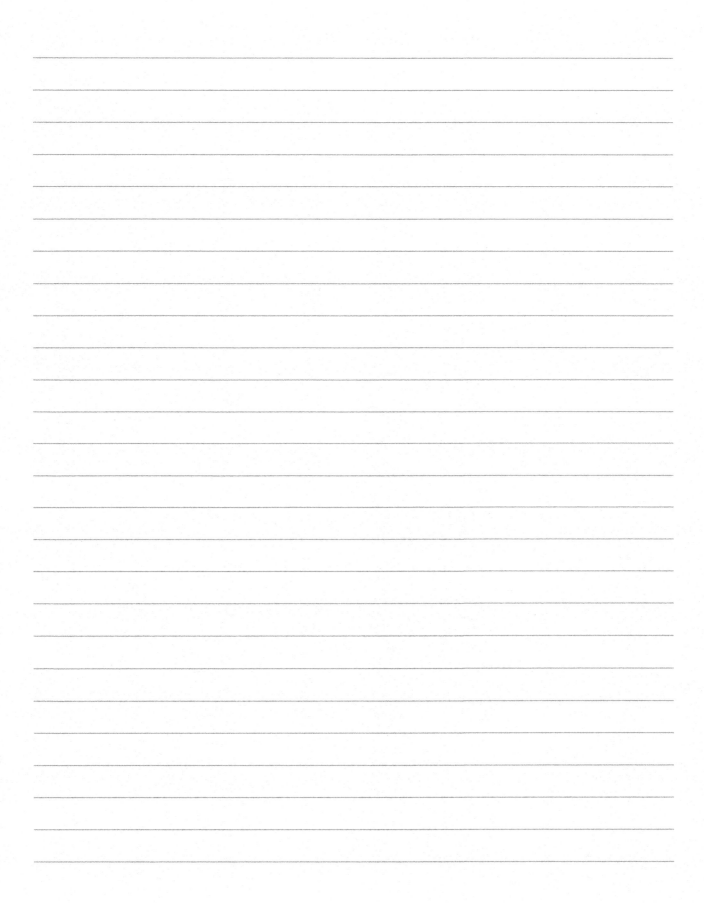

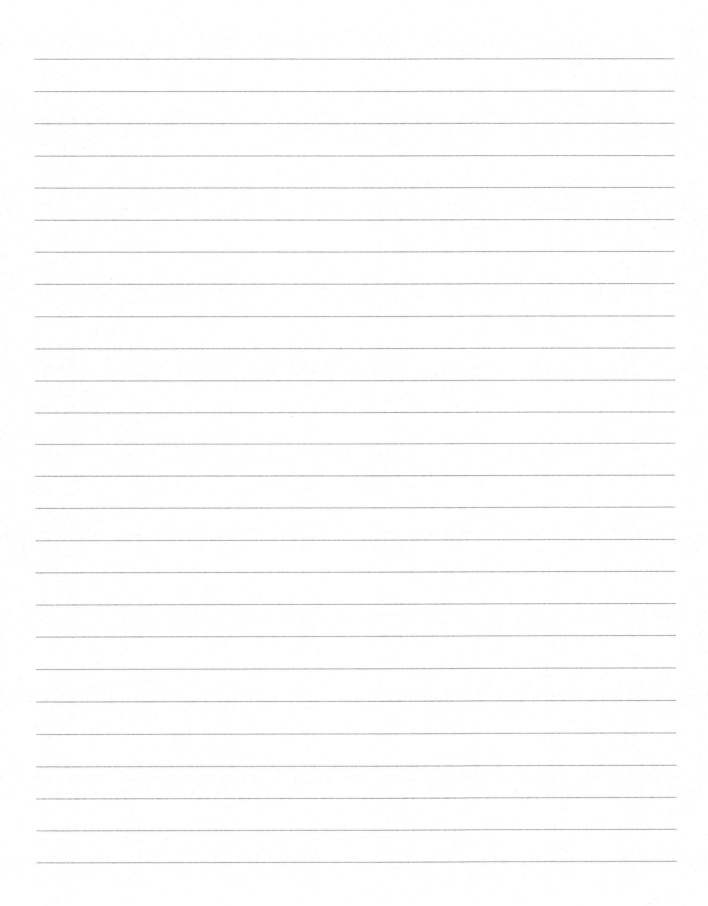

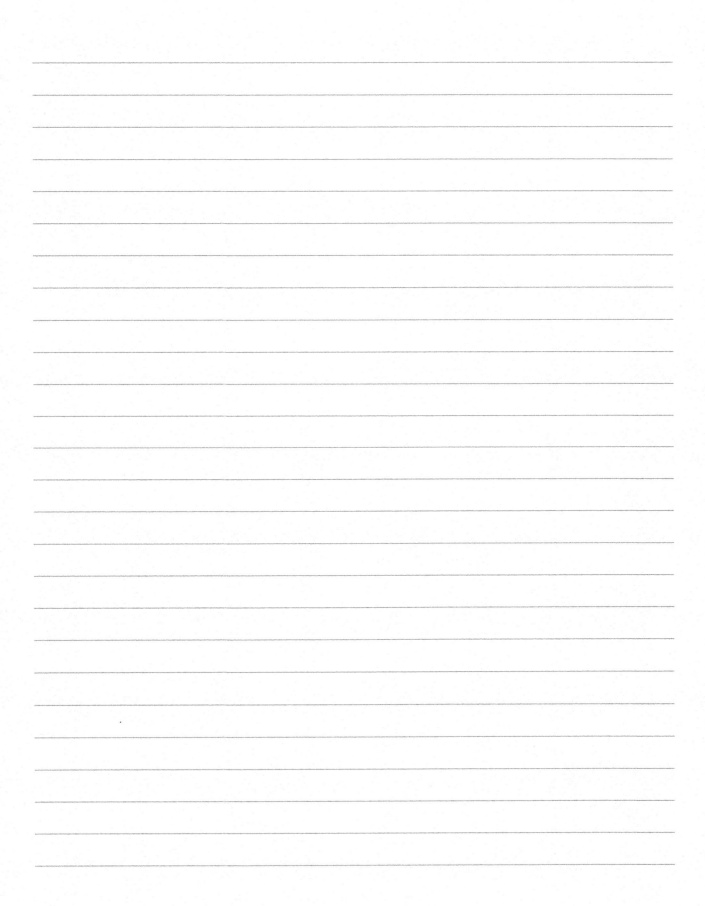

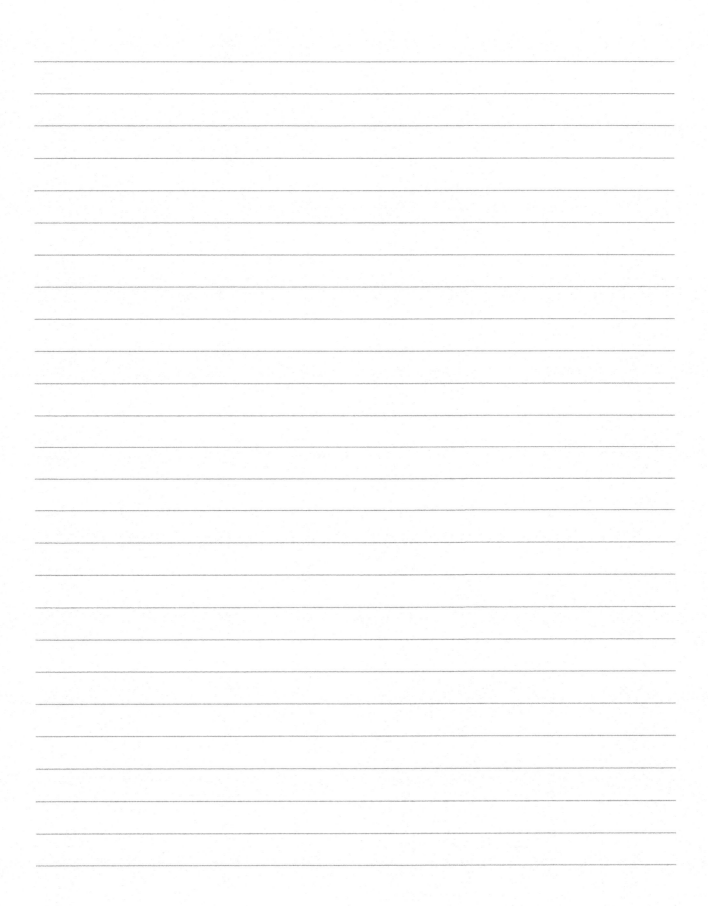

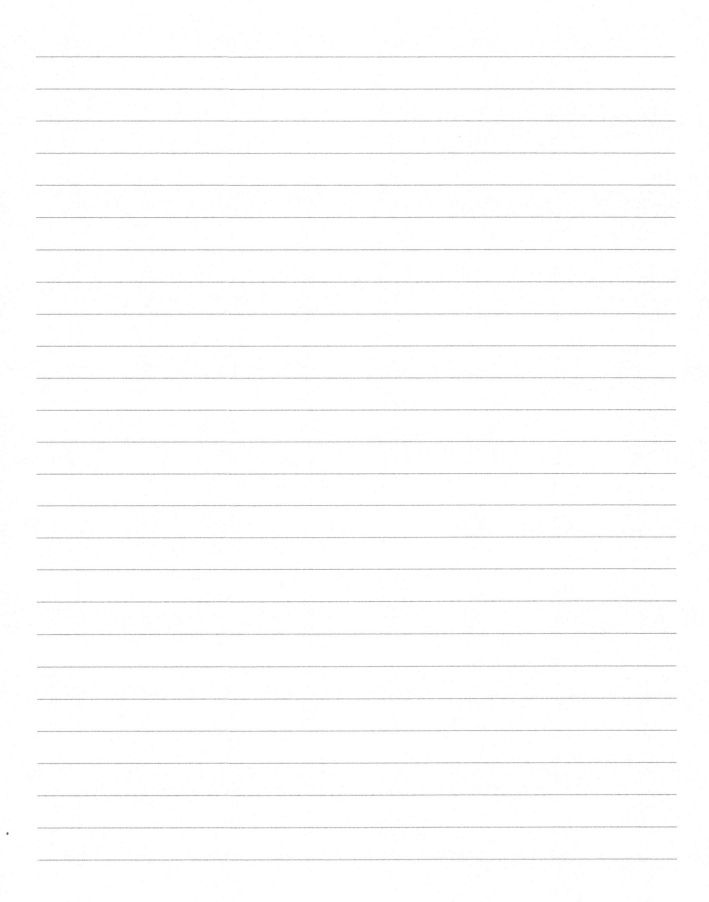

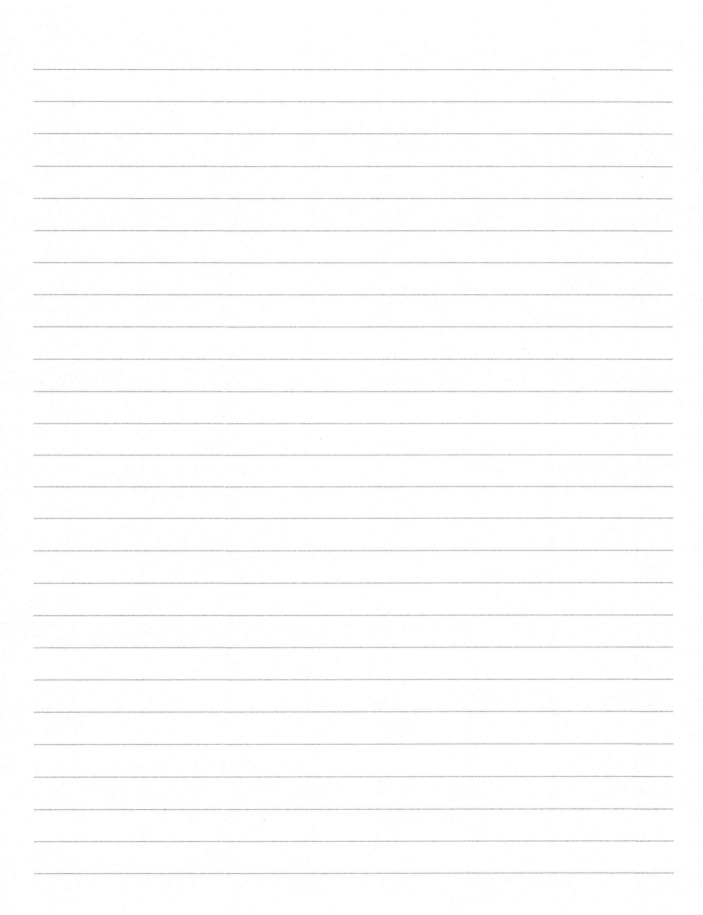

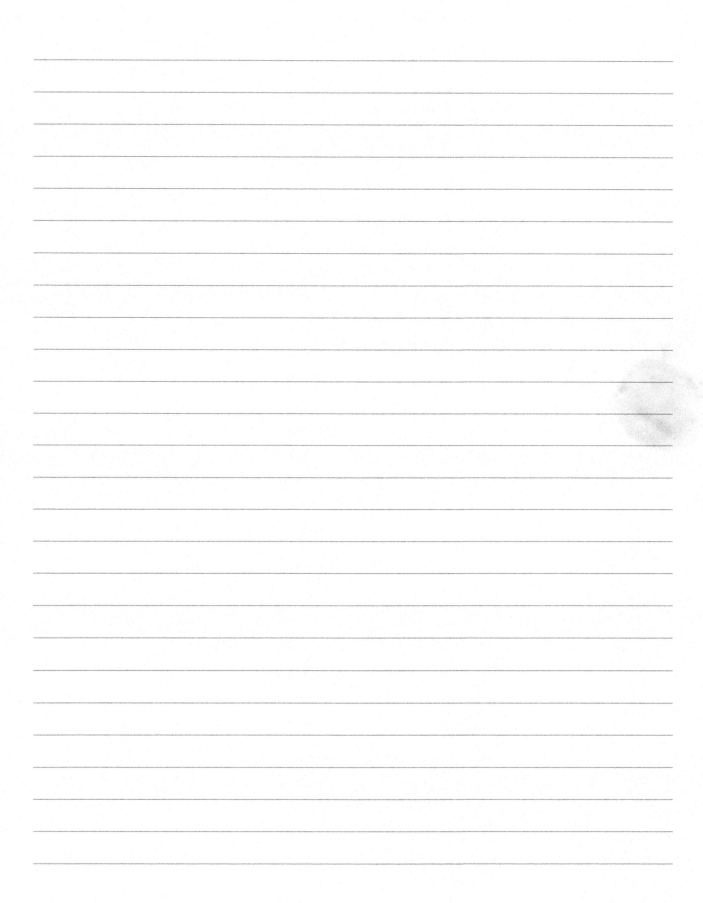

Made in the USA
Coppell, TX
23 January 2021

48690676R00063